MONSTER
COUNTING BOOK
1 to 20

Copyright © Frances Mackay 2021

The moral right of the author has been asserted.
All rights reserved. No part of this book may be reproduced or transmitted by any person or entity, including Internet search engines or retailers, in any form or any means, electronic or mechanical, including photocopying (except under statutory exceptions provisions of the Australian Copyright Act 1968), recording, scanning or by any information storage and retrieval system without prior written permission of Frances Mackay, author.

www.francesmackay.com

Design by Nicky Scott
www.nickyscottdesign.com
Illustrations supplied by Dreamstime

ISBN 978-0-646-85093-1

1

ONE green monster pulling a funny face.
What funny faces can you make?

2

TWO flying monsters.
How many pairs of wings do they have?

THREE weird monsters.
What colour is the monster with three eyes?

FOUR colourful monsters.
How many teeth do they have?

FIVE baby monsters.
Which monster has three legs?

SIX monster shadows.
Which shadow matches this monster?

7

Find SEVEN things that are different.

A big red monster with **EIGHT** eyes.
How many legs does it have?

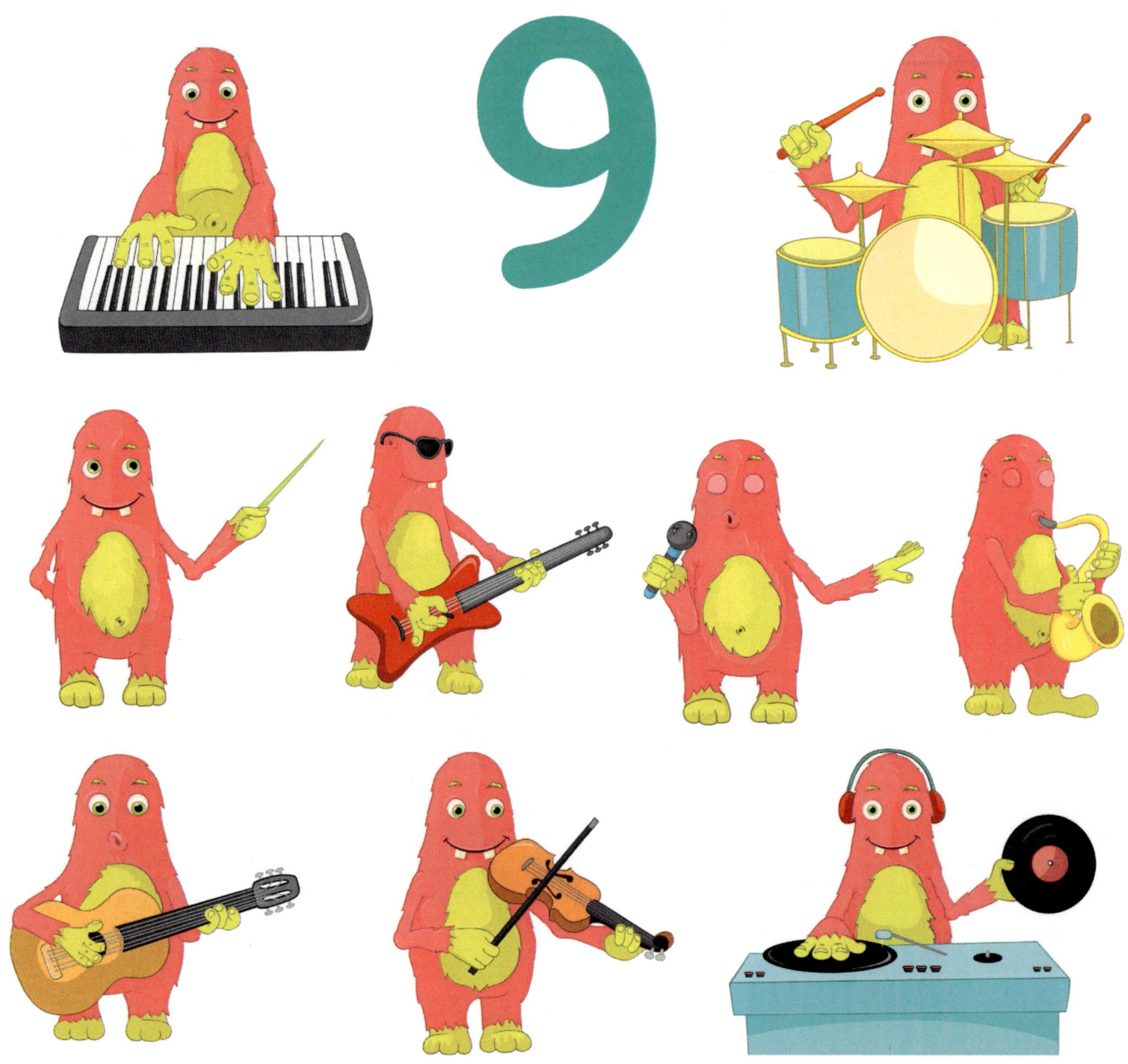

NINE monsters who love music.
How many are playing an instrument?

10

TEN monsters playing sport.

Can you name each sport?

ELEVEN blobby monsters.
How many monsters are orange?

12 **TWELVE** sad monsters.

How would you cheer them up?

13

A sneaky monster with **THIRTEEN** spines.
Count the spots.

14

This monster has **FOURTEEN** teeth.
How many fingers and toes?

15

FIFTEEN cute monsters.

Which one do you like best?

16 **SIXTEEN** funny monsters.

Which two are exactly the same?

17 SEVENTEEN balloons. Count them.

How many monsters at the party?

18

EIGHTEEN monsters dancing.
How many monsters are blue?

NINETEEN happy monsters.
How many have their eyes closed?

20

One blue monster fishing for stars.
There are **TWENTY** stars. Count them.

Now you can count from 1 to 20.

1 2 3 4 5 6 7
8 9 10 11 12
13 14 15 16
17 18 19 20

Well done!

Made in the USA
Monee, IL
22 July 2022